My name is

● ●

Oxford University Press, Great Clarendon Street, Oxford OX2 6DP

Oxford   New York
Athens Auckland Bangkok Bogotá Bombay
Buenos Aires Calcutta Cape Town Dar es Salaam
Delhi Florence Hong Kong Istanbul Karachi
Kuala Lumpur Madras Madrid Melbourne
Mexico City Nairobi Paris Singapore
Taipei Tokyo Toronto

and associated companies in
Berlin   Ibadan

Oxford is a trade mark of Oxford University Press
Illustrations copyright © Julie Park 1997
Text copyright © Oxford University Press 1997
First published 1997
1 3 5 7 9 10 8 6 4 2

A CIP catalogue record for this book is available from the British Library

ISBN 0-19-910417-4 (hardback)
ISBN 0-19-910418-2 (paperback)

Printed in Hong Kong by OUP Hong Kong

My first counting book

Illustrated by Julie Park

Consultant: Peter Patilla

Oxford University Press

2

two

3

three

4

four

5
five

6

six

1 2 3 4 5 6

7
seven

8

eight

9
nine

1 2 3 4 5 6 7 8 9

10
ten

1 2 3 4 5 6 7 8 9 10

13
thirteen

14

fourteen

16
sixteen

11 12 13 14 15 16

17
seventeen

18
eighteen

11 12 13 14 15 16 17 18

19
nineteen

11 12 13 14 15 16 17 18 19

Can you find these?

1 one orange orang-utan

2 two tumbling tigers

3 three thirsty toucans

4 four funny flamingos

5 five frisky foals

6 six skipping squirrels

7 seven stripy skunks

8 eight enormous elephants

9 nine nimble nanny-goats

10 ten talkative turkeys

11 eleven electric eels

12 twelve tough tortoises

13 thirteen tiny tadpoles

14 fourteen flying fish

15 fifteen frolicking frogs

16 sixteen spotty sea horses

17 seventeen smiling snails

18 eighteen energetic ewes

19 nineteen naughty newts

20 twenty tired tawny owls

Can you count?

Which numbers can you see?

Who is first?   Who is third?
Who is sixth?   Who will be last?